BrandVision Marketing Presents…

Digital Marketing Blueprint:

A Strategic Guide to

Achieving Measurable Results

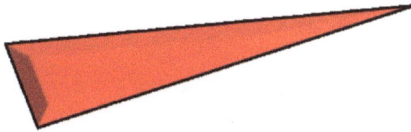

By Scott Trueblood

BVM Publishing

Library of Congress Control Number: 2025916706

Published by:

BVM Publishing * 8913 Town & Country Circle /#1077 * Knoxville, TN 37923

Author: Matthew Scott Trueblood

Printed and bound in the United States of America.

ISBN 978-0-9840665-0-6

Table of Contents

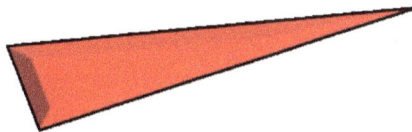

Preface

Welcome to *Digital Marketing Blueprint*—your strategic guide to building marketing that actually works in the modern age. Whether you're a corporate decision-maker or a curious beginner stepping into the digital world, this book offers a clear, proven roadmap to real, measurable results.

But before we dive into pixels and platforms, a few words about yours truly.

My friends call me "True." Maybe they're just too lazy to say my full name. Maybe they think it's edgy. Maybe it's because they caught a few too many episodes of *True Blood* and just can't un-hear it. (And for the record—no, I'm not a vampire. And yes, that show saved me from spelling my name everywhere I went.)

"True" is also a nickname I wear proudly—not just because I tell it like it is, but because I value everything such a nickname would encompass.

Yes, I realize that makes me sound like more of a collie than a real live dude, but fun fact: I grew up with collies. Amazing dogs. Loyal, smart, always ready for an adventure…just as I was, growing up in the 80's in rural Indiana. On the next page, you'll meet a few of my childhood companions—collies Captain and Friskie along with two of our cats, Tabby, and Twiki. (And yes, my sister too. She's the one holding Twiki.)

I'm just a regular guy from Southeastern Indiana who proudly calls Salem, Indiana my hometown. After high school, I made my way to Knoxville, Tennessee—home of the Smoky Mountains, the Tennessee Vols, and my second home for decades now.

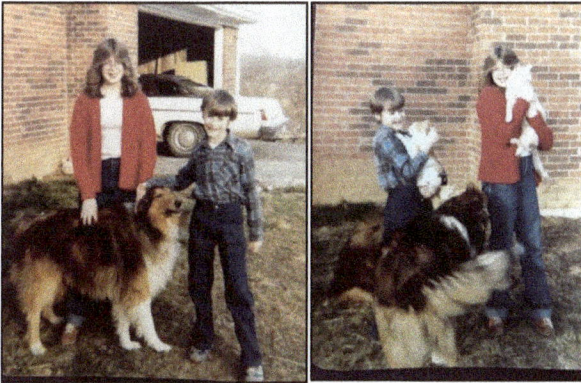

Things I love? My family (Hi Sis! Hi Puddin'! Hi Sauce! Rylee and Ella!), baseball (Go Yankees!), basketball and football (Go Vols!), and *Star Wars*—yep, full-on nerd status. I've had my share of wins and stumbles along the way. Which is to say: I'm human. Just like…well, most likely, you.

And no—I'm not a self-proclaimed marketing "guru" or "ninja." I come from a no-nonsense, humble Midwest upbringing where ego wasn't admired—it was politely ignored. So instead of claiming I'm the best, I'll just share what I know and let you be the judge.

A Bit About BrandVision Marketing

I started BrandVision Marketing in 1993—back when marketing meant radio, TV, newspapers, billboards, and the Yellow Pages. It began as a side hustle while I was selling radio ads at a local NewsTalk station. I wasn't a born salesperson, but I *loved* marketing. The thrill of connecting 'people with a problem' to 'businesses with a solution'? That was the good stuff.

What started as helping my radio clients with their broader strategy turned into a full-fledged agency by 1998. And while the media landscape has changed dramatically since then, one thing hasn't: **the importance of the Marketing Mix**.

Back in the '90s, the Marketing Mix meant reaching people across six key channels—and doing it often enough to stay top of mind. Today, the number of channels has exploded, but the mission is the same: reach your audience, and do it with consistency and purpose.

What changed the game?

The internet.

No media platform—*not* radio in the 1920s, not TV in the '30s—has had a greater cultural and business impact. Now, it's true…AI (Artificial Intelligence) may evolve to make a similar or even greater impact one day. But as of this writing the internet clearly holds that title. It has completely reshaped how consumers behave and how businesses must respond. Whether you're selling cars, gutters, insurance, healthcare, or legal services—your buyers are online. And that means your marketing better be there too.

So…What *Is* Digital Marketing?

Digital marketing includes all marketing efforts that use the internet or electronic devices—websites, search engines, social media, email, mobile apps, text messages, and more. But here's the thing: it's not about using *everything*. It's about using the *right* tools, at the *right* time, in the *right* way.

It's more than just being online. It's about building visibility, trust, and revenue at scale.

Thanks to digital, your market is no longer just your street, town, or zip code—it could be across the country or across the globe. But to compete, you need more than a website and a prayer. You need strategy.

At BrandVision Marketing, we've helped businesses across industries build that strategy—one that generates real growth, real leads, and real results. Now, we're handing that blueprint to you.

Let's cut through the noise, ditch the fluff, and focus on what *actually* drives growth.

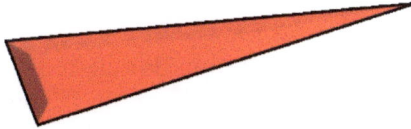

1

The Foundations of Digital Marketing

Let's be honest: when many first hear the term "digital marketing," they picture C-3PO and R2D2 peddling used Land Speeders (sorry, I couldn't resist!). Well…not so much—and no droids (yet). But it's not tech fluff—rather, it's about strategy that connects, converts, and gets results.

Digital marketing is more than simply posting online or running an ad. It's a cohesive system of strategies and tools designed to connect businesses with their audiences, create meaningful interactions, and drive conversions.

In the past, traditional marketing dominated—billboards, TV spots, print ads, radio and direct mail. Today, consumers are online, mobile, and empowered with information. Digital marketing lets you meet consumers where they are with personalized, timely, and measurable messages. Let's first look at the heart of it all.

Core Components of Digital Marketing:

- **Content Marketing** – Creating and distributing valuable content to attract and engage your audience. Your content lays the foundation for what your brand is all about. It attracts interested buyers and guides them through the sales funnel.
- **Search Engine Optimization (SEO)** – Optimizing your website and content to rank higher in search

engine results with the goal of driving more pertinent traffic to your site through those organic rankings, which presently still accounts for around 60 percent of clicks.

- **Email Marketing** – Nurturing leads and customers through targeted email communication. Email marketing when used as an opt-in with current customers and prospects, help you deliver sought after content while helping you build a foundation for a strong community of interested consumers.
- **Social Media Marketing** – Engaging users on platforms like Facebook, Instagram, LinkedIn, TikTok, and X (Twitter). There are few better ways for your brand to connect with consumers in a meaningful way than the use of social media. The key thing to remember is that it is 'social'—keeping sales pitches to the minimum and focus on engagement and relationship-building that furthers the brand to the consumer.
- **Paid Advertising (PPC or Pay-Per-Click)** – Running targeted ad campaigns on platforms like Google, Bing, Meta, and YouTube. This Digital Marketing strategy hits consumers who are very typically warm prospects, actively searching your product or service.
- **OTT/CTV (Over-the-Top/Connected TV)** -- OTT (Over-the-Top) advertising refers to the delivery of video content over the internet, bypassing traditional network, cable or satellite television. Meanwhile, CTV (Connected TV) advertising specifically targets ads displayed on internet-connected devices like smart TVs, streaming devices (i.e. Amazon's FireStick or Roku), as well as gaming consoles. OTT encompasses all devices that stream content over the internet, including smartphones and tablets. CTV's focus is on the big-

screen experience. It resembles traditional TV advertising in many respects, but is delivered through apps and streaming services. Both OTT and CTV offer unique experiences, with CTV often integrating more traditional ad formats during live TV or ad-supported apps. Both are highly targeted, capable of reaching consumers beyond mere demographic characteristics, but on psychographic interests as well. Targeting cigar smokers…luxury auto buyers…and so much more is easy.

- **Analytics** – Tracking performance, user behavior, and ROI to aid in making data-driven decisions that move your audience through the sales funnel and closer to your brand and a purchase.

Benefits of Digital Marketing:

- **Targeted Reach** –Digital Marketing allows you to reach your ideal customer with unmatched precision, using demographic, geographic, behavioral, and interest-based targeting. Unlike traditional media—such as radio, TV, or print—where audience insights are based on general assumptions (e.g., a NewsTalk radio listener vs. a Top 40 fan, or a sports viewer vs. a cooking show enthusiast), digital platforms offer far more advanced and granular targeting capabilities across multiple dimensions.
- **Measurability** – Unlike traditional channels, digital provides clear metrics: clicks, conversions, bounce rate, etc. With solid analytics, you can learn where you 'caught' the prospect; and possibly where you 'lost' them.
- **Affordability** – One of the best aspects of Digital Marketing for small-to-medium size businesses is the ability to build your campaign around flexible

budgets. You can start small and scale based on ROI.

- **Speed and Agility** – You can pivot campaigns in real-time. I remember days of picking up radio spots on reels from production studios. Need a change in that spot? Well, you're looking at days to re-produce. Digital Marketing allows flexible real-time pivots in campaign creative.
- **Personalization** – Use data to serve content tailored to each user's journey. With such fine-tuned targeted reach, you are able to deliver specific creative content to a specific audience. Such personalization strategies are great for brand-building—a true brand voice that drives sales.

The Digital Marketing Mindset

Successful digital marketing isn't about chasing trends. It's about strategic alignment—connecting your brand's purpose with your customer's needs, and delivering that value consistently.

Think beyond tactics. Ask:

- What do my customers need?
- Where do they spend time online?
- How do I guide them from interest to action?

This book will walk you through each of the digital pillars—from your website to analytics—and show you how to bring them together into one clear, growth-driving system.

Let's begin by understanding the new digital customer.

2

Understanding the Modern Buyer

Today, each of us generates and consumes more information than at any other time in human history. The average person processes up to 74 gigabytes of data every day—from TVs and smartphones to computers, tablets, billboards, and countless other sources. To put that into perspective: we now take in more data in a single day than someone living in the 15th century would encounter in an entire lifetime.

I grew up in the 1980s—Madonna on the radio, Lakers vs. Celtics on the court, big hair, and parachute pants everywhere. Those were good times filled with unforgettable memories. Back then, the average person processed only about 20% of the information taken in just three decades later. By 2011, Americans were consuming five times more information daily than they did in 1986— roughly the equivalent of reading 174 newspapers a day. In 2009 alone, the average American absorbed around 34 gigabytes of data each day, marking a staggering 350% increase in information intake over nearly 30 years.

Where does this information overload max out? Time will tell, but the point for our purposes is clear: Today's buyer has access to more information, more options, and more channels than ever before.

Today's consumers are not passive recipients of advertising—they're active researchers, comparison shoppers, and brand evaluators.

Understanding the psychology, behaviors, and expectations of this new customer is the foundation of an effective digital strategy.

Key Characteristics of the Modern Buyer:

- **Informed** – Today's buyers are more educated than ever. Before even reaching out to a salesperson, they've likely read reviews, watched videos, downloaded whitepapers, and compared competitors. They come to the conversation with specific questions, expectations, and a clear idea of what they want. This means brands must offer valuable, accurate content up front to stay relevant in the decision-making process.
- **Distracted** – Modern consumers are bombarded with information from every direction—social media, news, ads, notifications—all competing for their attention. As a result, attention spans have shortened dramatically. To cut through the noise, content must be immediately engaging, visually appealing, and delivered in bite-sized, easily digestible formats. Clarity, creativity, and brevity are no longer optional—they're essential.
- **Skeptical** – With so much misinformation online, buyers are naturally cautious and slow to trust.

Flashy marketing isn't enough; people want authenticity. Brands must prove themselves through transparency, consistent messaging, and third-party validation like reviews and testimonials. Building trust takes time, but it is truly the foundation of every successful relationship in the digital age.

- **Empowered** – Access to information, technology, and choices has shifted power to the buyer. They now expect a seamless, personalized experience—one that respects their preferences, remembers their behavior, and adapts in real time. Whether it's through AI-powered recommendations or custom content, businesses must deliver tailored solutions that make buyers feel seen, valued, and in control.
- **Multi-Channel** – Today's customer journey is far from linear. Buyers interact with brands across multiple touchpoints: websites, search engines, social media, email, apps, chatbots, and more. Each channel plays a role in shaping perceptions and influencing decisions. To succeed, brands must deliver a consistent and cohesive experience across all platforms—wherever the buyer chooses to engage.

Stages of the Digital Buying Journey:

1. **Awareness:** The buyer realizes they have a problem or need.
    - SEO, Paid Search, blogs, social media, and ads help them discover your brand.
2. **Consideration:** They evaluate different options.

- o Product pages, testimonials, comparison guides and webinars provide value.
3. **Decision:** They're ready to buy and choose a solution.
   - o CTAs (Call-to-Action), offers, reviews, and a seamless checkout process are key.
4. **Post-Purchase:** The experience continues.
   - o Email follow-ups, loyalty programs and great support help retain customers.

Using Buyer Personas

A persona is a semi-fictional representation of your ideal customer. Creating them helps you:

- Tailor content and messaging
- Choose the right platforms to present your marketing and branding messaging
- Understand objections and motivators

Example Buyer Persona:

- Name: Startup Sam
- Age: 32
- Role: Founder of a SaaS company
- Goals: Increase traffic and demo sign-ups
- Challenges: Limited time and marketing knowledge
- Preferred Platforms: LinkedIn, YouTube, email

With personas, you can speak directly to your audience instead of using vague, broad messaging. This allows you to identify Pain Points that a prospect is experiencing and address those directly through your marketing.

Meeting Buyer Expectations

- **Navigable websites** – Consumers need fast access to information. Load times matter. Remember, short attention spans are…ooh, a squirrel. Where was I? That's right. Short attention spans are part of the human experience now, so everything needs to be finger-tip accessible.
- **Mobile-first design** – Over 60% of buyers start their buying journey on a mobile device. According to Google, over 59% of shoppers say that being able to shop on mobile is important when deciding which brand or retailer to buy from. Salesforce reports that more than 60% of B2B buyers use their mobile devices to research products before making a purchase. In eCommerce, nearly 70% of website traffic comes from mobile devices, though conversions are often completed later on desktop. So while the percentage can vary slightly depending on the industry (B2C vs. B2B), it's safe to say that mobile is now the dominant starting point for most buyer journeys. *(See Chart 1 on Page 17)*
- **Authentic content** – Share behind-the-scenes, real customer stories. Remember your Buyer Persona and craft content that speaks to their Pain Point(s). At BrandVision Marketing, we use a lot of real customer testimonials in creating marketing content for clients. One of the keys is having those real customers discuss how they overcame their own Pain Points through their relationship with the client. Real customer stories strike a chord with other real customers. Use them in your own content.
- **Consistent experience** – From social media to email to your site, your brand should feel unified. Continuity is one of our staples of building a successful and profitable brand. Each touchpoint

should have the same branded look and feel as the last and next. Continuity within the customer's experience matters greatly to the overall marketing experience…to the digital marketing experience and greatly to the brand.

Conclusion

Understanding your buyer is not a one-time task—it is very much ongoing. The more you know the more precise and effective your marketing will be.

In the next chapter, we'll explore how to turn this understanding into offers that speak directly to your ideal customer's needs and desires.

CHART 1—Mobile Devices & Searches

Context	"Mobile First" Percentage
Online shopping occasions	> 60%
Retail product searches	77%
eCommerce website traffic	> 70%
B2B buying research	> 60%

Chapter 3: Crafting Offers That Convert

Let me save you a little heartache: You can run the best ad, nail the SEO, and have a funnel so smooth it practically sings—but if your offer doesn't make people say "Yes, please!"…it's all for nothing. Crickets!

Your offer needs to solve a real Pain Point to resonate and move the needle. Simply put: At the heart of every successful digital marketing campaign is an irresistible deal. Offers are more than just products or services—they are solutions to real problems. The goal is simple: Ease the consumer's Paint Point!

What Makes an Offer Irresistible?

A high-converting offer solves a specific problem, speaks directly to the customer's desires, and presents a clear, low-friction path to action. Keep that phrase in mind: 'low-friction path'. Today's consumer wants and expects few hoops to jump through…few obstacles to overcome…when making a purchase. If an offer is going to profitably convert to sales, it needs to be as low-friction as possible.

What makes an offer irresistible? Here is a list of five must-haves when accomplishing the goal of irresistibility:

- A clear **benefit** ("what's in it for me?")

- A sense of **urgency** ("why now?")
- **Social proof** (who else found it valuable?)
- **Risk reversal** (money-back guarantees, free trials)
- **Clarity** (no confusion about what's included)

Remember, you're addressing specific Pain Points being experienced by your target audience. Make sure that your offer includes each component in that checklist as it leads to "Problem solved!!!"

Types of Offers in Digital Marketing

Offers ideal for the Digital Marketing realm include the following:

- Lead Magnets
- Tripwire Offers
- Core Offers
- Upsells and Cross-Sells
- Continuity Offers

Lead Magnets: Free content offers (guides, checklists, and webinars) in exchange for contact information.

◆ *Example*: BrandVision Marketing offered a free "5 Rules for a Profitable Marketing Foundation" in exchange for an email address via a landing page.

✉ *Goal*: Build an email list by providing immediate value to potential leads.

Tripwire Offers: Low-cost offers designed to convert leads into buyers (e.g., $7 mini-courses).

♦ *Example:* A course creator promotes a $5 "Facebook Ads Mini-Course" to new subscribers who downloaded a free checklist.

▬ *Goal*: Turn cold leads into paying customers with minimal friction and start building buyer trust.

Core Offers: Your main product or service.

♦ *Example:* That same course creator sells a $197 comprehensive Facebook Ads Masterclass as the main product.

℧ *Goal*: Deliver your flagship service or product that drives most of your revenue.

Upsells and Cross-Sells: Additional purchases related to the core offer.

♦ *Example*: After purchasing the masterclass, buyers are offered a $99 one-on-one strategy session (upsell) or a $39 Instagram Ads toolkit (cross-sell).

📈 *Goal*: Increase average order value by offering relevant, high-perceived-value add-ons.

Continuity Offers: Subscriptions, memberships, or retainers.

♦ *Example*: Following a course purchase, customers are invited to join a $29/month membership group for ongoing ad updates, live Q&As, and peer support.

☐ *Goal*: Generate recurring revenue and long-term engagement with ongoing value delivery.

The Offer Stack

Rather than offering a single product, you can increase perceived value through an "offer stack." This is a bundle of bonuses, extras, or enhancements that sweeten the deal:

- Main Product ($X Value)
- Bonus #1 (e.g., exclusive training)
- Bonus #2 (e.g., downloadable templates)
- Guarantee or refund window
- Total value vs. actual price

This technique helps prospects feel like they're getting more than they're paying for. But please, folks…do NOT insult anyone's intelligence here. The ole, "$37,995.00 value can be yours for $29.95!!!" does nothing but create eye rolls galore…even among the most gullible consumers. Keep it real. Keep the offer attractive and a true win for the consumer. But keep them moving further and further into a deeper relationship and connection with your brand.

Crafting Offers with Buyer Psychology in Mind

People buy for emotional reasons and justify with logic.

Do not discount or dismiss the emotional element. Emotion lies at the heart of powerful marketing and lasting brand resonance. A brand that merely states its features misses the deeper connection that emotions foster—with authenticity, storytelling, and sensory triggers creating true engagement. Emotional branding taps into truly connective feelings like nostalgia, joy, and belonging—those intuitive, often subconscious drivers of loyalty. Consider Coca-Cola's "Share a Coke" campaign: by personalizing bottles with names, it moved beyond product promotion to spark joyful connections among friends and family. This kind of emotional bond—the "companionship or love" consumers feel toward a brand—translates into repeat business and word-of-mouth advocacy.

Sensory elements also play a big role: Starbucks' warm lighting, curated music, and signature aroma combine to produce comfort and familiarity, reinforcing emotional appeal at every visit.

Ultimately, brands that understand and evoke emotion—rather than rely solely on facts wrapped in features and benefits—forge deeper, more memorable relationships and stand out in a crowded marketplace. Logical processes then quickly follow to rationalize those emotional elements.

Tapping into these key psychological principles will help you connect with consumers, converting offers into cash:

- **Scarcity:** Limited spots, limited time, or limited quantity. FOMO is real. The FEAR OF MISSING OUT strikes at the heart of many, especially when it comes to an offer that can address a Paint Point. Limitations invoke that FOMO. Use scarcity legitimately and it can be a big winner for you.
- **Urgency:** Time-sensitive bonuses or discounts. The standard rule to follow is simple: put a deadline on every offer that is no longer than 10 days to two weeks. Anything more than that will be dismissed and forgotten.
- **Anchoring:** Show the original price before the discount. Combine a special price offer versus the original with **Scarcity** and **Urgency** and you have a certain winner.

- **Social Proof:** Showcase real results and testimonials from satisfied customers. Promote positive Review links.
- **Reciprocity:** Give value upfront to build goodwill and further the relationship. Remember, branding principles are still very much in play in the Digital Marketing space. Brands are all about relationships. Reciprocity can serve as a strong handshake to open the conversation and create that warm-spirited relationship with the prospect.

Optimizing Your Offer for Conversion

Here is a reality check, folks. You will not hit a grand slam every time. In fact, sometimes a seeing-eye single is more than welcome!

Sometimes your offer falls flat. It happens. Here's how to refine an offer that underperforms:

- **Reposition the value:** Are you highlighting outcomes, or just features? Address Pain Points!
- **Simplify the CTA (Call-to-Action):** Fewer steps = more conversions.
- **Add risk reversal:** Make it easier to say yes.
- **Test different headlines:** Sometimes one line can make the difference between sparking attention and being ignored.
- **Split test bonuses or price points:** Learn what resonates.

Offer Placement Matters

Where and how you present your offer affects conversions:

- **On your Website's homepage** with a clear call to action
- **In blog posts** as contextual CTAs
- **In pop-ups** based on user behavior
- **At the end of videos or webinars**
- **In email sequences** following nurturing content

Examples of High-Converting Offers

1. **Free SEO Audit Tool:** Great tool for BrandVision Marketing to start conversations.
2. **Exclusive Webinar:** With a time-sensitive replay.
3. **Mini-course:** For coaching or knowledge-based businesses.
4. **Discounted Bundle:** For eCommerce brands.
5. **VIP Access or Waitlists:** For SaaS or app launches.

Summary

If you want better results from your marketing, start by improving your offer. Craft something people truly want, present it clearly, reduce their friction, and watch conversions increase. In Chapter Four, we'll explore how your website plays a pivotal role in delivering and supporting these offers.

Chapter 4: Your Website – The Digital First Impression

Your website is often the first substantial interaction a potential customer has with your brand. It's your digital storefront, your credibility check, and your sales platform all rolled into one. If your website doesn't convert, it creates a gaping leak in your entire marketing funnel.

The Purpose of a Business Website

Different companies have different goals for their site. Obviously a site that sells kids clothes online will have different goals than a personal injury attorney or an accounting firm. However, a great website should:

- **Educate** the visitor about who you are and what you offer
- **Enhance and Further Your Brand**
- **Build trust** and establish authority in your niche
- **Guide the visitor** to take a specific action (e.g., buy, book an appointment, sign up for a newsletter)
- **Capture leads** to nurture in future marketing

Think of it not just as a digital brochure but as a tool for converting warm prospects into strong advocates.

The Elements of a High-Converting Website

1. **Clear Value Proposition:** Within seconds, visitors should know what you do, who this site is for, and why it matters.
2. **Strong CTA (Call to Action):** "Schedule a Call," "Get a Free Quote," "Start Your Free Trial"—make it prominent and consistent with your specific and unique branding elements throughout.
3. **Mobile Optimization:** Over 50% of traffic comes from mobile. If your site isn't mobile-friendly, you're losing potential business. Make sure your site is mobile friendly.
4. **Fast Load Time:** 1 in 4 visitors abandon sites that take more than 4 seconds to load. It's true. Patience is not a virtue among today's consumer. Make sure your site hits their device fast.
5. **Visual Hierarchy:** Use headings, colors, whitespace, and images strategically to guide the eye. Navigability is critical. Make sure these components help lead the consumer to finding the information they are seeking. More on this soon.
6. **Trust Builders:** Testimonials, client logos, security badges, and clear contact information should be present throughout.
7. **SEO-Ready Content:** Pages should include keywords, meta descriptions, and schema markup. All the critical components that help a site score highly in the organic rankings on major search engines should be present on-page and off-page.

Homepage Must-Haves

- A clear headline and sub-headline
- One or two CTAs above the fold
- A brief explanation of services/products

- Visual elements that reflect your brand
- Testimonials or social proof

Readability Meets Navigability

Not too many moons ago, Print Ads were a staple as a marketing resource. Size was vital. Placement was important. But ad design was incredibly crucial because industry research had shown us how consumers view ads—tracking their visual trail.

The eye began an 'ad read' near the left corner and tracked right…back left, etc. (See Image below) That same research holds true when considering a consumer's visual tracking of your website.

Start strong interest at the Top Left and track right…limiting the damage done in that dastardly Corner of Death. The ultimate goal is a user-friendly experience.

About Page Tips

People do business with people. An About page is far more than a company biography—it's a powerful connection tool that transforms strangers into advocates. This is an opportunity to share your brand's origin story by weaving in personal history, challenges overcome, and what led you to create your business. Doing so humanizes your brand and helps readers emotionally connect—not just understand what you do, but why you do it. That motivation is a critical brand bonding opportunity. It greatly furthers the relationship.

Use this space to highlight your mission and values, clearly articulating what guides your decisions and why customers should care. Research confirms that 58% of people align their purchases with brands whose values resonate—and this is exactly what your About page must showcase.

If relevant, introduce your team with real photos and bios. Putting faces behind the name boosts credibility and fosters trust—readers want to know who they're working with. Remember the value of familiarity within the marketing spectrum. Familiarity is vital!

Lastly, showcase media features or awards such as press mentions, certifications, or accolades. Highlight testimonials or success stories. All serve as positive trust signals and strengthen your authority in the eyes of visitors.

By weaving these elements together—storytelling, mission, team, and proof—you create an About page that builds genuine trust, fosters emotional engagement, and often converts visitors into customers.

In short, your company's About page provides an opportunity to:

- Share your story
- Highlight your mission and values
- Introduce your team (if applicable)
- Showcase media features or awards

Product/Service Pages That Sell

- **Focus on Benefits, Not Just Features...**Remember, it's about addressing Pain Points. Highlighting benefits is far more successful along those lines than a long list of features. A high -converting product page goes beyond listing specs—it shows users "what's in it for them." Highlight how a lightweight laptop improves portability or how a skincare serum restores confidence. When a feature is paired with its specific benefit, it all works together to address a Pain Point. This helps users visualize improvements that you can help them make to their lives.
- **Include a Compelling CTA...**Your Call-to- Action (CTA) should stand out—use bold, action- oriented text such as "Buy Now," "Order Today", "Get Started," or "Request an Estimate". Position it above the fold so that consumers do not need to scroll beyond their initial screen view. Make it stand out by using contrasting colors and consider sticky buttons on mobile. Shopify experts note

CTAs must be front and center to avoid friction and guide users toward purchase. Remember, no one wants obstacles. Ease of purchase is key to conversion.

- **Use Bullet Points for Readability…** Web users scan content rather than read line-by-line. Bullet points break complex information into bite-size chunks—use them for key features, specs, or benefits. Web reads should be quick-reads. People don't curl up with a great website on a Saturday night, nor are they deep-diving into *War and Peace*. Keep pages skimmable and reinforce your value proposition clearly. Bullet Points can help you accomplish that exact goal.

- **Add FAQs to Overcome Objections…** Anticipate customer doubts and address them upfront. A dedicated FAQ section can cover topics like sizing, ingredients, compatibility, shipping policies, and returns—removing friction and reducing support inquiries. Consumers insist on 'knowing what they're getting into' when engaging with your brand. FAQs help you head off a lot of objections in short order. Plus, such structured info significantly lowers bounce rates while building buyer confidence.

By focusing on user-centered benefits, prominent CTAs, a scan-friendly layout, and preemptive reassurance, you can transform your product or service pages into a trusted, high-converting digital salesperson.

In summary, your website is your brand's digital first impression. Be mindful of the following and you're on your way to Digital Marketing success:

- Focus on **benefits**, not just features
- Include a compelling CTA
- Use bullet points for readability
- Add FAQs to overcome objections

Blog for Authority & SEO (Search Engine Optimization)

A regularly updated blog supports your SEO strategy and builds authority. Publish content that:

- Answers customer questions
- Solves problems
- Positions your brand as their solution to a Pain Point
- Reflects your expertise

Landing Pages for Campaigns

When running ads, webinars, or lead magnets, use **dedicated landing pages** with a singular focus and no distractions. These convert higher than general website pages because they are focused solely on addressing a singular Pain Point.

Integrations That Matter

- **CRM integration:** Capture leads directly into your database
- **Live chat or chatbot:** Improve response time and user engagement, while enhancing the customer's brand experience
- **Analytics tools:** Track user behavior and conversions (e.g., Google Analytics, Hotjar)

Common Website Mistakes

You know what do to. Now, be mindful of the common mistakes that derail the goals of your website:

- Overly vague messaging that's not on-brand
- Cluttered design with no clear path
- Poor-quality images or stock photos
- Lack of clear CTAs on each page
- Outdated content or broken links

Summary

Your website is not a passive asset—it's an active marketing tool. When done right, it builds trust, educates, and guides users toward action.

In the next chapter, we'll show you how to amplify your website's reach through intentional, results-driven social media marketing.

5

Chapter 5: Social Media with Purpose

Social media is more than a megaphone for broadcasting content—it's a tool for connection, relationship building, and lead generation. When used with intention and purpose, platforms like LinkedIn, Instagram, Facebook, and even TikTok can drive significant business growth.

While nearly every business can benefit from a social media presence, the platform's effectiveness varies across industries. Social media tends to perform best in sectors where consumers are naturally inclined to engage— industries that inspire enthusiasm, lifestyle alignment, or community identity. For instance, a children's clothing boutique, a healthy snack brand, or a specialty cigar shop may generate vibrant followings because they spark interest, emotion, and even a sense of belonging. These brands often lend themselves to "fandom," where followers engage not just with the product, but with the lifestyle it represents.

By contrast, service-based professionals like dentists or attorneys may not evoke the same communal energy. That doesn't mean they should avoid social media—it simply means their strategy will look different. These professionals can still provide real value by sharing helpful tips, answering common questions, or demystifying their services to build trust.

As you plan your social media strategy, understand what your audience expects from your category. Not every business will go viral—and that's okay. Set realistic expectations and focus on delivering consistent, relevant content that supports your goals.

Choosing the Right Platforms

You don't need to be everywhere—you need to be where your audience lives. Where it already spends time online. For **B2B brands**, that typically means **LinkedIn**—the professional network where 84% of marketers find the most value—and increasingly, **YouTube** for sharing thought leadership videos, tutorials, and case studies. Brands like Salesforce…even B2C brands like Nike, Wendy's and Chipotle, have also seen success using engaging, "lo-fi" vertical videos, which are 'low-fidelity' videos typically shot on a smartphone with a DIY feel to spark meaningful connections.

For **B2C brands**, visually driven platforms like **Instagram**, **Facebook**, **TikTok**, and **Pinterest** are essential. Instagram's 1 billion users and highly visual format work well for storytelling and influencer partnerships. TikTok offers viral reach and Pinterest supports discovery-driven shopping behaviors, making them prime spots for lifestyle and retail brands.

If you're in a **niche market**, go where your community gathers—whether that's **Reddit**, **Discord**, or specialized **Facebook Groups**. On Reddit, you can engage in honest dialogue; on Discord, you can build close-knit communities. These channels foster deeper trust and loyalty, further enhancing the consumer's relationship with your brand.

It can all seem overwhelming and activity is vital. Create a content plan that coincides with your overall marketing efforts. If needed, just start with one or two platforms. Master on-brand consistent messaging, content formats, and engagement strategies on those platforms before expanding. Spreading yourself too thin across too many channels dilutes impact and dampens success. Focused effort builds stronger brand authority and higher ROI.

In summary:

- **B2B brands:** Focus on LinkedIn and YouTube
- **B2C brands:** Instagram, Facebook, TikTok, Pinterest, as well as YouTube
- **Niche markets:** Reddit, Discord, niche Facebook Groups

Start with 1-2 platforms and master them before expanding.

Define Your Social Media Goals

Social media can drive different types of value. Make sure your goals are well-defined and your strategic path clear. Define your goals as one or more of the following:

- **Brand awareness** – getting in front of your prospective consumers is a must
- **Lead generation** – create valuable lead magnets that will move prospects to your own database; participating in the social media world is playing in

someone else's toy box…they can change the rules at any time and take their toys and go home

- **Community building** – Increase followers to give your brand a louder voice with more prospects
- **Customer service** – Showcase your customer care programs by using social media as a direct way of handling issues or by sharing your customer service wins
- **Boost Engagement** – deepen the connection the consumer has with the brand
- **Driving website traffic** – move prospects away from the social side that is more about info-tainment and toward the more sales driven aspect…your website

Content Buckets for Engagement & Growth

Creating your content calendar is important. Remember, social media should be exactly that…social! Do not use a hard-sell approach too often. Promotion and Sales related posts should enter your content equation about ten percent of the time.

Most of your content should be either informational (about your industry and specific brand) or entertaining. Or ideally both! Making people laugh will never go out of style. Put a smile on someone's face repeatedly via your content and they will come back.

To maintain consistency and variety, group your posts into categories:

1. **Educational:** Tips, how-to, insights
2. **Entertaining:** Memes, stories, humor

3. **Inspirational:** Quotes, testimonials, case studies
4. **Promotional:** Offers, events, launches
5. **Personal:** Behind-the-scenes, team highlights

Use a simple content calendar to plan posts across buckets.

The Algorithm Loves Engagement

It's often joked that if you want to boost engagement on social media… just spell someone's name wrong. Seriously—send a LinkedIn message and call "Jonathan" by "Josh," and suddenly you've got his full attention. Turns out, egos are surprisingly interactive.

But all jokes aside, there's truth in the idea: platform algorithms love engagement. Whether it's a 'like', a 'comment', a 'share', or even a passive-aggressive correction, interactions tell the algorithm your content is worth showing to more people. So, the more you genuinely engage with your audience—ask questions, reply to comments, celebrate your followers—the more visibility you earn. In short: real interaction isn't just good manners, it is good marketing.

Try to increase engagement by trying some of the following:

- Asking questions
- Starting conversations in comments
- Using polls and quizzes
- Encouraging user-generated content

Respond to every comment and DM to build relationships. Even if the comment seems to be AI generated, reply. The engagement will get a favorable nod by the algorithm, increasing potential exposure and reach other consumers.

Hashtags & Captions

Hashtags and Captions serve valuable roles within social media for user and operator alike. A well-crafted hashtag helps you organize and categorize content. For users, it helps them locate pertinent information. Both can help you expand your reach, while bringing specific context to your posting.

Here are some helpful tips on Hashtags and Captions:

- Use niche hashtags (#MarketingForCoaches, not just #marketing)
- Include a clear call to action (CTA): "Comment below," "Tag a friend," "Download the guide"
- Make your captions scannable with line breaks and emojis

Leverage Stories, Reels, and Video

Social Media is a great resource when connecting with your target audience. When the goal is relationship-building, the medium literally rocks! Nothing allows for the relationship building that truly brings brands to life like stories.

Use the power of storytelling through your content whenever possible. Short-form video is dominating attention. Use Reels, Stories, and TikToks to:

- Demo your product or service
- Show customer testimonials
- Share tips in bite-size form
- Tell stories about your brand journey

You don't necessarily need high-end production. Authenticity wins.

Paid Social Media Strategy

It took more than three years after the launch of FaceSmash—yes, that was Facebook's original name, straight out of a college dorm room—before Mark Zuckerberg and crew realized, "Hey, maybe this thing can make money!" Thus, Facebook Ads were born... and so was the modern era of paid social media marketing.

Unlike traditional media that targets broad demographic groups (gender, age, zip code), paid social ads let you zoom in on people's interests, behaviors, and even quirks. Want to reach knitting lovers for your quilting expo? Done. Got a new protein powder for fitness buffs who also follow goat yoga? Facebook probably knows exactly who they are—and what they ate for breakfast.

But proceed with caution. Paid ads can be a digital gold mine... or a black hole for your budget. One unfortunate marketer we know burned through $30,000 on Facebook Ads without a single lead. That's not strategy—that's a cautionary tale.

To avoid becoming a sad meme, always start with a clear objective. Know your audience—where they hang out, how they scroll, and what makes them click. And since organic reach fell off a cliff somewhere around 2018, pair your organic content with smart, targeted paid campaigns.

Start small, test often, and keep a sharp eye on ROAS (Return on Ad Spend). Paid social isn't magic—but when done right, it can feel pretty close.

In Summary:

Since algorithms shifted years ago and organic reach has declined. Complement your organic content with paid social:

- Boost top-performing organic posts
- Run retargeting ads for warm audiences
- Use lead generation campaigns (e.g., free checklist or webinar)

Start small, test regularly, and focus on ROAS (return on ad spend).

Track the Right Metrics

Don't get lost in vanity metrics (likes, follows). After all, no one ever asks their waiter, "Do you take Visa, Mastercard or Likes?" Simply put. 'Likes' are great, but do not translate directly into cash! Push the ole ego of vanity metrics aside and instead, monitor:

- Engagement rate
- Click-through rate (CTR)
- Conversion rate from social traffic
- Follower growth in target demographics
- Direct messages and lead inquiries

Social Media Mistakes to Avoid

Yes, Social Media can very much be a gold mine, but do not let it become fool's gold to your marketing objectives. Avoid the following pitfalls:

- Posting without strategy or consistency
- Overusing a Hard-Sell approach and not educating or entertaining
- Ignoring comments and DMs
- Not analyzing performance
- Jumping on every trend without relevance

Summary

Social media works best when you treat it as a relationship-building channel—not just a digital billboard. It's not about shouting the loudest or posting the most. It's about showing up with purpose, engaging with sincerity, and offering content that feels helpful, relevant, or even entertaining. Be intentional with your messaging—every post should have a reason to exist. Be consistent in your presence—trust is built over time, not in a one-off campaign.

When you focus on building real connections instead of chasing vanity metrics, the results become far more meaningful. Social media is a long game. The likes, comments, shares, and saves are just signals—what truly matters is whether your content builds trust, encourages dialogue, and reflects the values of your brand. Your audience doesn't expect perfection; they want authenticity, responsiveness, and value. Show up with that, and you'll earn not just attention, but loyalty. Next up…email!

Chapter 6: Email Marketing That Nurtures & Converts

Email marketing remains one of the most cost-effective and impactful tools in your digital strategy. It builds relationships, nurtures leads, and drives consistent conversions—if done right.

Note that phrase…'if done right'. That is critical to a successful use of most all marketing mediums but especially true with Email Marketing.

Cold Email vs. Subscriber Lists: Who's Really Happy to See You in Their Inbox?

Picture this: You're at a party. One person walks up and says, "Hey! I loved your playlist last week—got any new recommendations?"

That's your subscriber.

Then another person appears out of nowhere and says, "Hi, you don't know me, but I'd love 60 seconds of your time to talk about cloud-based invoice automation."

That's your cold email.

See the difference?

Cold email has its place in digital marketing—but let's be honest, it's a slow burn. A very. Slow. Burn. Unless you're a big brand like Apple or Beyoncé (or, ideally, both), most people aren't thrilled to hear from a stranger trying to sell them something. Cold email is like knocking on someone's door without an invitation and hoping, they not only answer, but also buy a vacuum cleaner.

It's not impossible to succeed with cold email—especially if you're targeted, personalized, and respectful—but it does require time, patience, and more nurturing than a bonsai tree. You're building trust from scratch. In an inbox filled with 117 other cold pitches…and two passive-aggressive emails from Karen in accounting…that's no easy task.

Now let's talk **subscriber lists**—your email marketing goldmine. These are people who *chose* to hear from you. They signed up. They said, "Sure, I'm into what you're doing." They may not be super fans yet, but at least they're warm leads. You're not fighting for their attention so much as reminding them why they liked you in the first place.

Here's a stat that drives it home: **Email marketing to subscribers delivers an average ROI of $36 for every $1 spent**, according to Litmus. That's not just good—it's borderline magical. With a subscriber list, your emails are less *"Who is this???"* and more *"Oh hey, I remember them!"* And that recognition leads to higher open rates, better click-throughs, and yes—actual conversions.

The Bottom Line?
Cold email can be effective, but it's a long game—and not one for the impatient. Email marketing to subscribers, on the other hand, is like dating someone who already likes your taste in coffee. You've got a great and potentially productive head start. So while you *can* build an empire

one cold email at a time, it's a lot faster (and frankly more fun) to build relationships with the people who have already said "yes."

And in digital marketing, a warm lead is always better than a cold shoulder.

Why Email Still Works

Email delivers directly to your audience without relying on social media algorithms. With high ROI (remember…averaging $36 for every $1 spent…WOW!), it offers incredible leverage when properly personalized and automated.

Types of Email Campaigns

1. **Welcome Series:** The welcome series is like a digital handshake—your chance to say, "Hi, we're glad you're here!" Use it to introduce your brand's story, values, and voice. Share your top-performing content or helpful resources to make a strong first impression. A well-crafted welcome sequence sets expectations and builds early trust, increasing the chances that your new subscriber sticks around for the long haul (and soon becomes a customer).
2. **Nurture Sequences:** Think of nurture sequences as the slow dance of email marketing. You're not selling—you're warming people up. This is where you deliver consistent value: insightful blog posts, quick tips, how-to guides, case studies, or even short personal stories. It's about positioning your brand as helpful, knowledgeable, and trustworthy. Over time, you'll move people from curious

subscribers to confident buyers—without ever feeling pushy or salesy.

3. **Promotional Emails:** Promotional emails are your chance to spotlight the goods—sales, limited-time offers, product launches, or seasonal discounts. The key is to make them timely, relevant, and irresistible. Include strong subject lines, bold calls-to-action, and a sense of urgency ("Only 24 hours left!" still works like magic). When balanced with value-based emails, promotions feel like helpful opportunities—not annoying sales pitches. Unlike social media, where your goals are relationship-driven and not delivering hard sell pitches, email marketing gives you the chance to be far more aggressive. Just don't overdo it. When aggressive crosses the line towards pushy, people will unsubscribe faster than you can say "flash sale."

4. **Behavioral Emails:** These are the mind readers of email marketing—automated messages triggered by what your subscribers do (or don't do). Someone abandons their cart? Remind them! Maybe you can throw in a discount or Free Shipping. Clicked a product link but didn't buy? Send them more info. Signed up for a webinar? Confirm it and tease what they'll learn. Behavioral emails are timely, personal, and wildly effective because they meet people exactly where they are in their journey.

5. **Re-engagement Campaigns:** Every list has a few ghosts—subscribers who haven't opened an email in months. Instead of writing them off, try a re-engagement campaign. Remind them of the value you offer, tempt them with an exclusive offer, or simply ask if they still want to hear from you. It's a smart way to clean your list, boost engagement, and potentially revive interest. Sometimes, all it takes is a clever subject line to bring a ghost back to life.

Building an Email List

Building a subscriber list is a daunting task. Maybe the consumer is a big fan already. Maybe you have a brand that sparks great enthusiasm. If so, you, my friend are way ahead of the game. If you don't…and yes, I feel your pain, you have to be a little more creative. If the latter, is indeed the case, you have to up the ante a bit. Offer a hearty hook or perky incentive to motivate someone to click that subscribe button.

Don't be duped, however, but don't be alarmed. It's a challenge but it's not impossible. Motivating a consumer to part with their email knowing it means more and more 'stuff' hitting the ole inbox is a budding task today. People are more guarded about their inboxes than ever before.

The key is value…offer an incentive that has a high-perceived value and your list will grow. Start with the following:

- Offer a free downloadable guide or checklist. Information with a high-perceived value is gold in this setting; and quite often worth a subscribe
- Free Webinars—especially if the webinar is able to be reviewed at the consumer's leisure
- Offer discounts and digital coupons with strong CTAs. (Call-to-Actions)
- Avoid buying lists—permission-based marketing wins for small-to-medium sized businesses…unless your brand's name recognition is off the charts in your market

Writing Emails That Get Opened and Clicked

- **Subject lines:** Your subject line is your first (and sometimes only) shot at grabbing attention. Be clear about what's inside, spark curiosity, or promise value—then back it up. Use power words, numbers, or questions (e.g., "Feeling stuck? Try this 5-minute fix"). A/B testing subject lines can also reveal what resonates most with your audience.
- **First line:** The first line of your email often appears in the preview text alongside the subject line—so it matters a lot. With that in mind, it serves as a strong motivator to open the message. Think of it as your hook. Make it compelling, relevant, or even a bit intriguing to entice the reader to click. Avoid wasting it on "Hi there" or filler text—go straight to the good stuff. Most likely, this and the Subject Line will be read, so make sure both are relevant and spark interest
- **Body copy:** The main message should feel like a friendly conversation, not a sales pitch. Write the way your audience talks, keep paragraphs short, and use formatting (like bullet points or bold text) to make it easy to scan. Focus on one clear idea per email to avoid overwhelming or confusing the reader.
- **CTA:** Every great email needs a next step. Whether it's "Book a free consultation," "Download this Time-Saving Checklist," or "Shop the Sale," make your CTA clear and actionable. Place it where it's easy to find—ideally more than once in longer emails. A strong CTA turns casual readers into active, engaged leads or customers.

Winning Subject Line Examples:

"Struggling to get traffic from Instagram? Here are 3 quick fixes to triple your…" Addressing a specific issue with a fast solution gets attention and can serve as a powerful subject line.

"You're missing out on this [free tool/strategy]" This plays on *FOMO* (Fear of Missing Out), which is a powerful psychological trigger. When people feel like they're being left behind or missing something valuable, they're more likely to click.

"How we boosted our conversions by 47% (and how you can too)" This one delivers *specific value* and uses numbers to create credibility. Subject lines with stats tend to perform well because they sound actionable, real, and results-driven.

Segment Your List

One-size-fits-all doesn't work. Segment based on:

- Demographics (job title, location)
- Behavior (clicked links, downloads)
- Lifecycle stage (new lead vs. past customer)

What I love most about Email Marketing is that campaign segmentations lead to much higher conversions in the short term. For example, you send out an email about a specific product and find that 37 percent of your opens clicked the link to that product! Great news, right? Even better. Now, you can put those 37 percent in a 'bucket' and create a segmented campaign hyper-focused on that product to move the consumer along the sales funnel.

Automate With Purpose, but Keep it Real

Use automation tools that keep communication alive and well, but do so in a very real way. Don't patronize. Connect.

- Trigger welcome sequences
- Deliver pertinent lead magnets
- Remind prospects about abandoned carts
- Onboard new clients

Automations save time and increase conversions, but be genuine with the content. Focus on building the relationship and furthering the brand in the mind of the consumer.

Email Is a Great Medium—When You Don't Use It Like a Maniac

Email marketing is powerful—*when* it's used well. Unfortunately, too many businesses use it… *well*, badly. Let's face it: email already comes with baggage. Between phishing scams and shady promos, it's no surprise people are wary. So when your message lands in someone's inbox, you're starting with a bit of a black eye.

And then some brands make it worse.

It reminds me of the street preachers at the University of Tennessee who used to set up camp and shout condemnations at passing students. I remember thinking, "You're yelling at people who…the majority of which… probably already share your faith—READ THE ROOM!" The same rule applies to email.

Those pushy follow-ups like "Wanted to Bump this Up!" or "I am waiting for your reply!" from someone I've never heard of? Straight to the trash bin. Or worse—unsubscribe, block, and a serious eye-roll, typically only aimed at your older sibling (unless you have a younger sibling, of course).

So, dear email marketers: **READ. THE. ROOM.**
You're a stranger in their inbox. Act accordingly. Lead with empathy, spark curiosity, and—most importantly—always be moving toward addressing a *real* Pain Point.

Want to stay out of email jail? Avoid these classic pitfalls:

- Over-emailing—or going silent for months
- Writing like a hype machine instead of a human
- Ignoring subject line and CTA testing
- Skipping deliverability basics (seriously, verify your domain)
- Hoarding inactive subscribers like digital skeletons

Track These Email Metrics

There's no doubt. Metrics matter. When it comes to Email Marketing, those analytics offer a great way to set objectives…follow the consumers buying path, and so much more. Set reasonable goals. Assess. And tweak as needed. Pay close attention to the following:

- **Open Rate:** % of recipients who opened the email
- **Click-Through Rate (CTR):** % who clicked a link
- **Conversion Rate:** % who took the desired action
- **Bounce Rate:** % of undelivered emails
- **Unsubscribe Rate:** High rates may indicate frequency or relevance issues

Summary

Email is about relationships. So is Email Marketing. When you prioritize value over promotion, use personalization, and align with the buyer journey, your emails become assets—not annoyances.

In the next chapter, we'll unlock how to run paid ads that actually work and don't drain your budget.

7

Chapter 7: Paid Ads That Don't Waste Your Budget (or Your Sanity)

When it comes to sales funnels, the closer your customer is to the bottom, the closer you are to hearing that sweet cha-ching. And guess what? **Paid ads are one of the fastest ways to show up right when someone's reaching for their wallet.**

Digital marketing gives us the power to meet potential buyers at every stage of their journey—from the vaguely curious to the "shut up and take my money" crowd. But when you're looking for the warm-to-hot leads (think: tropical vibes, SPF 30 kind of warm), paid advertising can be your VIP pass to the people already circling your checkout page.

PPC-You-At the Cash Register...

Here's where PPC (pay-per-click) comes in. Want to catch someone *literally* Googling your service or product category? PPC campaigns let you target those laser-specific keywords so your brand pops up at exactly the right time—*while they're actively searching* for what you offer. That's not interrupting their day; that's showing up right on cue.

But don't just throw money at the internet and hope for the best. **Keyword strategy is everything.** Some terms are hyper-competitive and cost more than your streaming subscriptions combined. Others are cheaper, niche-oriented, and still pack a punch.

Run a few breakeven scenarios before you dive in—best case, worst case, and "what the heck happened" case. Paid ads can absolutely drive growth, but only if you're playing smart and thinking long game. Too many businesses burn through ad dollars like it is Monopoly money, with nothing to show but a sad little click-through rate.

So before you hit "launch," ask yourself:

- Do I know who I'm targeting?
- Are my keywords dialed in?
- Do I have a solid landing page to catch the click?
- And, most importantly…
  Do I have a plan, or am I just feeding the algorithm and hoping it likes me?

Because throwing money at ads without strategy is like buying drinks for strangers and hoping one of them proposes. There's a better way.

Why Use Paid Ads? With Intent in Mind...

If paid ads are your rocket fuel, then search intent is your navigation system. Whether someone's typing "buy electric bike near me" (hot, transactional) or "benefits of electric bikes" (cooler, research mode), knowing their intent means you can meet them on the right runway.

Paid ads help you:

- Generate traffic and leads fast, especially from intent-rich searches
- Test your offers and audiences based on what users actually want right now
- Retarget past visitors who showed interest but didn't commit
- Scale profitable campaigns, once intent + offer + audience align

Choosing the Right Platform

Google Ads

Ideal for **high-intent** or *Transactional* searches—people actively looking to buy or sign up. Target "buy [product]" or "[service] near me" and show up before they even scroll.

Facebook & Instagram Ads

Perfect for **Informational/Consideration** intent—warm-up phase. Your ad interrupts a scroll with a story or offer. Later, retarget those who click, and watch them move into purchase intent.

LinkedIn Ads

Works best for **Commercial/B2B intent**—people researching vendors, like "best CRM for SaaS." Reach decision-makers in evaluation mode with your offer front and center.

Tip: Always ask—*Where is my audience in their 'intent' journey?* Then match platform, message, and offer accordingly.

Pick your platform based on your audience and funnel stage.

Ad Objectives That Align With Goals

Set clear goals that sync with a searcher's intent:
- Awareness (Informational): Use video views or impressions to capture a curious audience.
- Consideration (Commercial): Drive clicks or landing page views from those doing product research.
- Conversion (Transactional): Target buyers ready to pull the trigger—optimize for signups, purchases, and/or requests.

Creating High-Converting Ads

Great ads don't just look good—they speak directly to what someone wants. This is vital when the goal is to address a Pain Point...position your brand as the solution to that Pain Point and prompting a click to your site or Landing Page.

Create ads that focus on the following:

- **Hook:** Reflect their intent. "Want to buy [product]? We've got a deal."
- **Offer:** Match intent with value—free trial, discount, comparison guide.
- **CTA:** Use action words relevant to intent—"Buy now," "Get a quote," "Download guide."

Example: "Searching for 'best CRM for startups'? Get our free comparison guide today."

"Struggling with low website traffic? Download our free 7-day traffic boost plan now."

Use high-quality visuals or videos that match your message, while speaking to your brand.

Landing Pages That Keep Their Promises (Seriously, Don't Be That Marketer)

Let's be blunt: **mismatched landing pages kill conversions.** If someone clicks on a "Buy Now" ad and lands on a generic blog post or homepage, you've just confused—and likely lost—a perfectly good prospect.

Worse yet? **Making a big promise in the ad and then ghosting it completely on the landing page.**

Confession time: I ran a little mini digital marketing test this week. Just for kicks (and research, of course). I responded to two ads:

1. A car insurance company claiming *massive* savings for low-mileage drivers.
2. A "government program" offering free gutter replacement. Intriguing, right?

Both had strong hooks. I was curious.

Here's what happened:

- The car insurance quotes were **way higher** than what I'm already paying. No mention of low-mileage discounts...**AT ALL!** When I mentioned this to the rep that reached out they were obviously oblivious.
- The gutter folks? Looked at me sideways when I asked about that "government program." But hey, they *were* happy to quote me a full-price gutter system.

Long story short? **Bait-and-switch marketing is the fastest way to lose trust** (and get blocked, reported, or memed).

So please—**if your ad promises something, your landing page really needs to deliver.** Echo the same message. Reinforce the same value. Guide the visitor seamlessly to the next step with **no** surprises, confusion, or shady vanishing acts.

Because when you make an offer that resonates... you've earned their click. Don't waste it.

The point is: people are responding to an offer that resonated...do not send them somewhere where that offer is completely ghosted.

Instead:

- Mirror their intent: transactional visitors get checkout-style pages; Research-phase attendees get guides or comparisons.
- Include a single CTA that aligns: checkout, download, or book a demo—no distractions.
- Add trust elements like testimonials, reviews, or stats to validate intent.

Retargeting = Second Chances

Only a small percentage of consumers convert on the first visit. Don't be discouraged. It just reinforces the vitality of your brand and the solidity of your overall marketing process. Further, search intent evolves—someone in research mode today might buy tomorrow.

Retargeting is a marketing tactic using online ads to target prospects that have already raised their hand to your brand. They have previously visited or interacted with a website or social media presence. Retargeting allows your brand to serve tailored ads to potential customers based on their prior engagement. The goal is to effectively recapture their attention and increase the chances of conversion.

This strategy is particularly useful for re-engaging users who left a website without completing a purchase.

Use retargeting to:

- Show ads to people who visited but didn't convert
- Remind cart abandoners they still want the item

- Re-engage research-phase visitors with "ready to buy?" or "Have questions?" messaging
- Adapt tone based on intent: gentle nudge vs. hard sell

Retargeting boosts ROI because it capitalizes on evolving intent.

Avoid These Paid Ad Pitfalls

- Skipping audience research
- Sending traffic to poor-quality landing pages
- Not testing multiple versions of ads (A/B testing)
- Focusing only on clicks, not conversions
- Letting underperforming campaigns run too long

Track These Paid Ad Metrics

- **CTR (Click-Through Rate):** Indicator of ad relevance
- **CPC (Cost Per Click):** Shows cost efficiency
- **Conversion Rate:** How many clicks led to action
- **ROAS (Return on Ad Spend):** Total revenue / ad spend
- **ROMI (Return on Marketing Investment):** Look at the big picture across your marketing spectrum
- **Frequency:** How often your ad is seen (keep awareness high, but watch for ad fatigue)

Next-Level: Mastering Search Intent with Keyword Segmentation & AI

If you've ever felt like you're throwing spaghetti at the wall with your ad keywords and hoping something sticks—this part is for you. Let's talk about how to use **keyword segmentation** and **AI-powered intent analysis** to turn that spaghetti into a Michelin-starred strategy.

⚬ *Keyword Segmentation: Because "One Size Fits All" Never Fits Anyone*

Not all keywords are created equal. Some signal buyers in heat, others… just browsers in pajamas (btw…there's nothing wrong with surfin' in your pj's!).

Segment keywords into intent buckets:

- **Transactional**: These scream *"Take my money!"*
  Examples: "buy standing desk," "sign up for CRM trial," "best price Nike shoes"
- **Commercial Investigation**: The shopper who's *almost* there.
  Examples: "top rated accounting software," "Shopify vs. WooCommerce"
- **Informational**: The researcher. Interested, but needs nurturing.
  Examples: "how to run Facebook ads," "what is a business line of credit," or "what is a heloc"
- **Navigational**: Brand-specific or direct lookups.
  Examples: "Amazon login," "Nike return policy"

➡ ☐ **Pro tip**: Create separate campaigns (or at least ad groups) for each intent category so you can tailor your copy, bids, and landing pages accordingly.

🔍 *Search Term Reports: Hidden Gold Mines*

In your ad dashboard (Google Ads, Meta Ads, etc.), dig into **search term reports** to see what people actually typed before clicking your ad. It's like eavesdropping… but ethical.

What to look for:

- Irrelevant terms that wasted clicks (pause those suckers)
- Hidden gems—long-tail phrases that show serious intent (e.g., "Nike basketball shoes for outdoor")
- Common patterns (e.g., "best for beginners," "under $100") to inform future copy

Use these insights to refine:

- Keywords and negative keywords
- Ad copy (mirror the language people use)
- Landing pages (match their mindset)

🔍 *How AI Predicts Search Intent (Yes, It's Cool)*

Today's AI tools don't just help you guess intent—they **analyze it at scale**:

- **Natural Language Processing (NLP)** helps platforms like Google interpret subtle differences in queries. For example, "best yoga mat for bad knees" carries *intent + concern*, and AI knows it.

- **Predictive analytics tools** (like Clearscope, and MarketMuse) cluster keywords by intent and show what types of content or ads work best for each cluster.
- **Chatbot or CRM AI integrations** (like Drift, HubSpot AI) can even adjust messaging in real time based on user behavior and past search patterns.

Bottom line: AI can help you anticipate what someone wants *before* they fully articulate it. That's next-level targeting.

Bringing It Home

Pairing paid ads with **search intent** and **smart segmentation** transforms your campaigns from hit-or-miss to highly relevant. The closer you get to what your audience *actually* wants, the more they click, convert, and stick around.

So—ditch the guesswork. Let intent lead the way, and let AI help read the room.

Tips for Intent-Optimized Campaigns

- Build campaigns around **intent themes**—transactional vs. informational keywords. There is a difference and honing in on that can be the difference between attracting shoppers and salesmen
- Use long-tail, high-intent keywords and match landing pages accordingly

- Analyze search term reports to refine keyword and copy intent
- Segment audiences by intent and personalize messages for each stage

Tips for Overall Better Results

- Use lookalike audiences to find similar prospects
- Regularly refresh ad creatives
- Set up proper tracking (pixels, UTM links)
- Test small before scaling

Summary

Paid advertising can accelerate your growth when you align audience, messaging, and funnel stage. Start with small, test-driven campaigns and build from what's working.

If paid ads are your engine, then **search intent** is your steering wheel. Understand what searchers want—right now—and match your ad, platform, and landing page to that moment. When intent, message, and offer align, you're not wasting budget—you're making magic.

In the next chapter, we'll dive into how SEO and content can grow your business without upping your ad spend.

Chapter 8: SEO and Content That Compounds

Search engine optimization (SEO) and content marketing are powerful long-term tools that can compound growth without ongoing ad spend. When done right, these strategies work hand-in-hand to drive traffic, build trust, and establish your brand as a thought leader.

Why SEO Matters

You know the question you bantered around in 3rd grade: "What super power would you want?" Inevitably 'invisibility' entered the convo at some point, right? Sure, many just wanted to sneak into their locker room of choice for a peak; others put the local bank vault in the ole crosshairs. Regardless, of your chosen super power, invisibility is not something to shoot for when it comes to your online presence.

Most online journeys begin with a search. It is how consumerism has evolved over the years. Sure, some categories are 'referral'-charged, but even those typically go online for validation and verification before taking the next step.

If your business isn't showing up in those coveted organic rankings on search engines such as Google, Bing, Yahoo or Yandex to name a smidge few, you're invisible to potential customers. SEO helps with the following:

- Increase organic website traffic
- Capture high-intent leads
- Lower customer acquisition costs over time
- Build authority and trust with search engines and users

Core Elements of SEO

1. **Technical SEO**
   - Fast-loading website
   - Mobile-friendly design
   - Secure (HTTPS)
   - Structured data and crawlable site architecture
2. **On-Page SEO**
   - Keyword research and placement
   - Clean URL structures
   - Headings (H1, H2) for structure
   - Meta titles and descriptions
   - Internal linking to related content
3. **Off-Page SEO**
   - Backlinks from reputable websites
   - Guest posts, PR features, and digital partnerships
   - Directory listings and citations

Content Marketing: The Fuel for SEO

Content is what search engines rank—plain and simple. Back when SEO first took off, everyone scrambled to 'beat' Google with tricks like keyword stuffing and shady link-building schemes. But here's the truth: Google isn't out to get you. Its goal has always been the same—to connect people with the most helpful, relevant information.

So if you focus on creating genuinely useful content, you're already ahead of the game. The more value you provide, the more trust and traffic you'll earn.

And what qualifies as 'helpful'? Ask your prospects. Listen to your customers. Then answer their real questions with real insight. That's how you win at content. Here are a few types of content that perform well:

- Blog posts that answer common questions
- How-to guides and tutorials
- Video content (embedded on pages and on YouTube)
- Infographics and visual content
- Case studies and industry insights

Consistency is key. Simply put: Plan! A content calendar ensures you're regularly adding SEO-optimized content that aligns with your keywords and audience.

Keyword Research Basics

Good content starts with good keyword strategy:

- Use free online tools like Google Keyword Planner, to provide you with a vast database of keywords and content suggestions that can help you address topics to boost SEO and help your business rank better in search engines
- Target a mix of short-tail (high volume) and long-tail (specific) keywords--Short tail keywords are

search terms comprised of 1-3 words. They refer to very broad topics rather than specific ones. For example, "running shoes" is an example of a short tail keyword; whereas a long-tail keyword drills down…for example, "best running shoes for summer"

- Include local modifiers if location matters (e.g., "plumber in Chicago")—Our SEO offerings at BrandVision Marketing have always been broken into 'LOCAL' plans vs. 'NATIONAL' plans. For example, an online retailer selling cigars across the globe will have a different keyword strategy than a plumber seeking to dominate only in their backyard. To rank highly use local identifiers with a 'your town, state' next to your keyword is vital (e.g., 'busted pipe repair Louisville KY).

Finally, and this is important: Avoid keyword stuffing. Aim for natural usage that enhances readability. Remember Google's chief purpose (connect people with helpful info) and align with it. Stuffing keywords to the point of creating an unreadable page will not help you long-term, crawl after crawl.

Optimizing Existing Content

You don't always have to start from scratch. Revisit older pages and blog posts to:

- Update statistics or data
- Improve headings and structure
- Add keywords naturally
- Refresh images and visuals
- Add internal links to newer content

Measure Your SEO Success

One of the biggest mistakes I see business owners make is assuming that just launching a website earns them a front-row seat on page one of Google. Okay. Okay. Bearer of bad news here, but that's like setting up your beach towel and expecting the ocean to come to you. The reality? Around 90% of content gets zero traffic from Google *(source: Ahrefs)*.

The odds are, you're stepping into a keyword arena where competitors have been flexing their SEO muscles for years. So yeah—they've got a head start. Stay plugged into where you rank for your key terms, and build a strategy that helps you climb the ladder—one optimized step at a time. Do so by focusing on the following SEO metrics:

- **Organic Traffic** – Visits from search engines
- **Keyword Rankings** – How high you rank for targeted terms
- **Bounce Rate & Time on Page** – Engagement indicators
- **Backlink Profile** – Quantity and quality of inbound links

Use tools like Google Analytics and Google Search Console to monitor progress. Or request a **free** SEO Audit from BrandVision Marketing by visiting our website at: www.brandvisionmarketing.com

Common SEO & Content Mistakes

- Ignoring technical SEO issues (slow site, broken links, on-page and off-page issues)
- Writing for search engines instead of real people
- Not optimizing for mobile
- Publishing inconsistent or low-value content
- Failing to promote published content (social, email, etc.)

Content Promotion Tips

Your work doesn't end when you hit publish. Your blogging, vlogging and other content efforts are meant for eyeballs. Promote your content through:

- Email newsletters
- Social media sharing (organic & paid)
- Partner collaborations
- Syndication on Medium, LinkedIn, or industry sites

Summary

SEO and content marketing build a powerful inbound engine that continues to bring traffic and leads long after publication. By focusing on quality, strategy, and consistency, your content becomes an asset that compounds over time. Up next? Exploring how data and analytics turn guesswork into confident marketing decisions.

Chapter 9: Data that Drives Real Results

Data is your compass in the digital marketing realm. Without it, you're navigating blindly. Data is essentially your guide. Without it, you're basically sailing without a map—maybe even without a boat. So yes, data is vital. But let me be perfectly clear. There is only one metric that matters: Sales.

Sales are the only metric that truly matters.

You might crush an Aaron Judge-style homer deep into the left field seats, but if you just stroll back to the dugout instead of running the bases, guess what? No run tallied on the scoreboard. Same goes for markcting. Impressive stats are fine, but if they don't move the revenue needle, they're just distraction and, ultimately, noise.

That said, data is incredibly valuable—when used right. From quick wins that fill the cart to slow burns that simmer through your funnel (like that weekly email you send with care), data helps you make smarter decisions, boost performance, and avoid wasting your valuable budget.

Think of it as your secret weapon—less about vanity metrics, more about clarity, efficiency, and ultimately...profit.

In this chapter, we'll cover how to use data to refine strategies, improve results, and make informed decisions that support your business goals.

Why Data Matters

Every digital marketing action—from emails and ads to blog posts and landing pages—generates valuable data. But here's the catch: collecting data isn't the same as using it. It's like stockpiling ingredients without ever cooking the meal. The magic happens when you analyze and act on what the numbers are telling you. When used wisely, data becomes your GPS in the digital landscape, pointing you toward smarter decisions, better results, and bigger wins.

Here's how:

- **Identify What's Working (and What's Not):** Data shows you which strategies are gaining traction and which ones are falling flat. Maybe your ad gets tons of clicks but barely any conversions—it's all about Sales, right! Data helps you spot that and tweak accordingly before wasting more budget.
- **Understand Your Audience Better:** Are your customers mostly mobile users? Do they respond better to video than blogs? Are they bouncing from a specific page? Data reveals behaviors, preferences, and patterns so you can meet your audience where they are—and give them more of what they want.
- **Optimize Spend and Resources:** Marketing budgets aren't infinite (sad, but true...or I'd be on a beach at Turks & Caicos sipping a blackberry mojito...okay, guzzling a blackberry mojito), and

guessing is crazy expensive…if not just plain crazy. Data tells you where to double down and where to scale back. It helps you stretch your dollars and focus your energy on tactics that drive ROI, not just noise.

- **Accountability--Prove ROI to Stakeholders:** Whether it's your boss, your board, or your own bank account—at some point, you'll need to show that what you're doing is working. Solid data gives you the evidence to back up your strategy, justify your spend, and make the case for future investment.

Case in Point:

One client we worked with was spending heavily on Facebook ads. The strategy was bringing in a lot of traffic but very few sales. (Remember, it's the only metric that matters…time to tweak!) By digging into the data, we discovered that while the ads were doing their job, the landing page was not; and was primarily responsible for a high bounce rate. In other words, we built it…they came, but left way too quickly to notch traction. After a redesign of the landing page, conversion rates jumped by nearly 40%, and the client saw a double-digit savings in cost per acquisition (by more than 20%).

Without data, this fix would've been a shot in the dark— and a costly one at that.

Case in Point…The Sequel

Once upon a time in the land of Email Marketing (that's right, the sequels never live up to the originals…exception *The Empire Strikes Back*, of course) We began working with a client who utilized this medium extensively. When auditing their weekly email newsletter that had been humming along to a steady database of subscribers for than a year, we saw open rates hovering around 10% and click-throughs under 2%. Neither was anything to write home about. But the client was fairly pleased with those numbers as long as people were commenting that they 'saw it'.

A quick data dive showed the subject lines were too vague ("Weekly Update") and the key content was buried several scrolls down. Worse, mobile formatting was busted, which is where over 65% of their audience was opening.

We tested punchier subject lines, renamed the newsletter and moved the key Call-to-Action above the fold. Result? Open rates doubled, click-throughs tripled, and more importantly—revenue from email more than doubled.

Moral of both stories? Data told us what the audience wasn't saying out loud. Then we listened…adjusted…tweaked—and made it work.

Core Marketing Metrics to Track

When it comes to data and marketing metrics, it is way too easy to get bogged down in numbers…many of them meaningless at that! Focus on the following metrics to drive your goals toward Sales.

1. ⚙ Traffic Metrics

These help you understand how many people are visiting your site and where they're coming from.

- **Total Website Visits**: The total number of visits to your site (these numbers typically includes new and repeat visitors).
- **Traffic Sources**: Shows where visitors came from:
    - *Organic* (search engines),
    - *Direct* (typed URL or bookmarked),
    - *Referral* (from another site),
    - *Paid* (ads).
- **New vs. Returning Users**: Tells you how many visitors are brand new vs. how many have been there before—both are important for your marketing efforts.

---

2. ⚔ Engagement Metrics

These show how people interact with your website and content. These are metrics that help you determine the traction your brand is gaining in terms of building that invaluable lasting connection.

- **Bounce Rate**: The % of visitors who leave after only seeing one page. A high bounce rate may mean your content isn't matching what they expected.
- **Average Time on Page**: The average time a visitor spends on a specific page. Longer = more engaged.
- **Pages per Session**: The average number of pages a user visits in one session.
- **Click-Through Rate (CTR)**: The % of people who clicked on a link or ad compared to those who saw it. Great for measuring the appeal of your content.

---

3. ♂ Conversion Metrics

These focus on how well your marketing efforts turn visitors into leads or customers.

- **Leads Generated**: The number of potential customers (form fills, calls, signups, etc.).
- **Conversion Rate by Channel**: The % of visitors from a particular source (e.g. email, paid ads) who took a desired action.
- **Cost Per Lead (CPL)**: How much you're spending to acquire each lead. Lower is usually better.
- **Return on Ad Spend (ROAS)**: The revenue earned for every dollar spent on ads. (Example: ROAS of 4 means you earned $4 for every $1 spent.)

---

4. ♂ Retention Metrics

These help measure how well you keep customers coming back. The more they return, the deeper the bond with the brand!

- **Email Open and Click Rates**: The % of recipients who open your email and then click on something inside it.
- **Repeat Purchases or Visits**: Tracks how often someone comes back to your site or buys from you again.
- **Customer Lifetime Value (CLV)**: An estimate of how much a customer will spend with you over the long haul. High CLV = loyal, profitable customers.

✒ *Essential Tools for Tracking Digital Marketing Data*

• Google Analytics 4 (GA4):
The Swiss Army knife of website data. Track who's visiting, where they're coming from, what pages they love (or bounce from), and what actions they take. *Pro tip: Set up conversion goals to see what's actually driving Sales.*

• Meta Ads Manager (Facebook/Instagram):
For paid ads, this tool gives you the nitty-gritty on clicks, reach, engagement, cost per result, and more. Use A/B tests to refine creatives, headlines, and targeting.

• Google Search Console:
Want to know how your site performs in search? This one tells you what keywords are bringing in traffic, how your pages rank, and whether you're showing up—or slipping into oblivion.

• Email Marketing Platforms:
Typical email marketing platforms provide delivery rates, open rates, click-throughs, and even automation data. Look at what content your audience actually clicks—and adjust accordingly.

• CRM Software (like HubSpot, Zoho, or Salesforce):
Combine sales and marketing insights to see which leads are converting and where they came from. This helps align sales and marketing teams around shared data.

• Hotjar or Microsoft Clarity:
Want to see what users are doing *on* your site? These tools offer heatmaps, scroll maps, and even session recordings. It's like watching customers shop in real time (—minus the awkward eye contact).

📊 Quick Metrics Dashboard: Track What Matters Most

Category	Metric to Watch	Why It Matters	Tool to Use
☐ Website Traffic	- Sessions - Top Pages - Bounce Rate	See how people find you and where they drop off	Google Analytics
🎯 Conversions	- Leads - Purchases - Goal Completions	Measures real business results (not just vanity metrics)	Google Analytics, CRM tools
📧 Email Marketing	- Open Rate - Click-Through Rate (CTR)	Gauge subject line & content performance	Email Platforms (BrandVision Marketing, Mailchimp, etc.)
🔊 Paid Ads	- Click-Through Rate (CTR) - Cost Per Click (CPC) - ROAS	Keeps your budget lean and ROI high	Google Ads, Meta Ads
🔍 SEO Performance	- Keyword Rankings - Clicks from Search	Understand how you're showing up in search	Google Search Console

Starter Checklist: Are You Tracking...?

If you're just getting started and need a jumping off point...follow the list below. It will provide a soft, but useful landing for your first leap:

☑☐ Traffic & engagement on your top 5 pages
☑☐ Conversions (form fills, sign-ups, purchases)
☑☐ Email opens & clicks
☑☐ Ad performance (CTR, CPC, ROI)
☑☐ Keyword rankings for top 10 target terms

Setting Up KPIs

A KPI (Key Performance Indicator) is a **measurable value** tied to a business goal. A KPI demonstrates how effectively you are achieving key business objectives. KPIs are used to evaluate success across the board...within the company's walls and also in comparison to competitors. KPIs provide clear and quantifiable metrics to track progress, identify strategic areas for improvement, and inform decision-making moving forward as well as assessment and accountability upon review.

Examples of KPIs:

- Increase Sales in first quarter by 10% year-over-year
- Increase website traffic by 20% in 6 months
- Lower cost-per-click (CPC) by 15% in the next 1st quarter campaign and 17% in the 2nd quarter campaign
- Grow email list by 1,000 qualified leads this quarter

Make KPIs SMART:

- **Specific** – What exactly are you measuring?
- **Measurable** – Can it be tracked numerically?
- **Achievable** – Is it realistic based on your resources?
- **Relevant** – Does it align with your business goals?
- **Time-bound** – When do you want to achieve it?

So You Think You've Got a Winner: A/B Testing for Optimization

Sure, your campaign idea wowed everyone in the boardroom. Maybe your significant other swears it's the next viral sensation. I've been there.

One long-time client came to me positively *giddy* about a "guaranteed smash hit" ad concept. "My girlfriend came up with it—she's a marketing genius," he said.

"Fantastic!" I replied. "What agency is she with? Maybe I should try to poach her!"

"Well, technically, she's not in marketing," he said. "She's a server."

Now, there's absolutely nothing wrong with waiting tables—my first job was stocking Nike's and Converse. I get it. But I suggested we test the idea before we start popping champagne. Just in case.

That's where A/B testing comes in.

Split testing lets you run two (or more) versions of something—an ad, email, landing page—to see which one your audience actually responds to. Think of it as a friendly duel between concepts… where the data decides the victor.

Some examples:

- Email A uses humor and a punchy subject line. Email B strikes at the emotions.
- Landing Page A has a video. Page B uses a bold image.
- Ad A targets Gen Z. Ad B talks to Boomers.

One golden rule: **test one variable at a time**. That way, you know *exactly* what made the difference—was it the headline, the photo, or the fact that your dog was in the ad?

(Spoiler alert: it's always the dog…most likely a collie!)

Avoiding Vanity Metrics

Not all metrics are equally useful. In fact, many metrics…especially those in play within social media are there to boost ego and create that hopeful hug that never quite brings the warmth. Ignore those. They will rarely help. Instead focus on metrics that steer…if not drive…revenue. Think: conversions and CLV, for instance. Further, 'Likes', 'Shares' and 'Impressions' can create a valuable 'tell' when it comes to your digital marketing goals. However, unless they are tied directly to business objectives, do **not** let yourself get hung up on them. They're more about vanity than results.

Marketing Dashboards for Clarity

Using dashboards to visualize data in one place is key for the solo-preneur wearing five hats as well as for the marketing team wearing one hat apiece. There are tools that provide access for users to scope out data and coordinate their efforts. At BrandVision Marketing, clients get access to a dashboard that allows them to follow progress on everything from their SEO and PPC campaigns to their SMS and Social Media programs. Many such tools exist.

Tools like:

- Google Looker Studio (formerly Data Studio)
- HubSpot dashboards
- Custom Excel or Google Sheets

These help stakeholders and teams stay aligned on what matters most.

Summary

Data allows you to market smarter, not harder. It provides the insight needed to double down on what's working and cut what isn't. When interpreted correctly, data shifts marketing from a cost center to a predictable growth engine. And remember…ultimately, the main metric: Sales!

Next, we'll wrap up our journey with a step-by-step guide to building your digital strategy blueprint.

10

You've explored the core pillars of digital marketing—now it's time to pull everything together into a clear, focused game plan.

Think of this as your **Digital Blueprint**: a living, breathing document that guides your marketing decisions. It should be strategic but flexible, aligned with your brand identity and revenue goals, and ready to evolve as your business grows.

Let's walk through the process—**step-by-step**—with these 8 foundational moves to help you level up and lead in the digital space.

Step 1: Clarify Your Goals

Where are you now? And where do you want to be?

Answering those two questions is your starting line. The more honest and specific you are here, the more precise and profitable your strategy will become.

Start by asking:

- **What does success look like?** (Sales? Visibility? Sign-ups? Brand awareness?)
- **How will we measure that success?** (Use **SMART goals** and define your **KPIs**.)
- **Who is our target audience?**

- **What problems do we solve? What Pain Points do we address?**
- **How does our audience currently perceive us?** (Trusted name or total unknown?)

Define these clearly before making a single post, ad, or email send.

Step 2: Conduct a Digital Audit

Time to evaluate where you stand. Before you race forward, check under the hood and assess what's working—and what's not.

Use this quick audit to identify strengths, gaps, and golden opportunities:

- **Website:** Is it mobile-friendly, fast-loading, and easy to navigate? Does it guide users toward action?
- **Traditional Media:** What offline channels (Radio, TV, Direct Mail, Print, Outdoor) are you using— and do they align with your digital messaging? **Continuity matters across all touchpoints.**
- **SEO:** Are you targeting the right keywords? How are you ranking for them today?
- **Social Media:** Are your platforms active, consistent, and engaging? Are you showing up where your audience hangs out—or wasting time elsewhere?
- **Email:** Are you regularly nurturing leads with relevant content? Are your campaigns segmented to meet users at different stages of the buying journey?

- **Paid Ads:** Are you seeing ROI? Are your ads targeting ready-to-buy consumers or just curious browsers? (**Hint:** search intent matters!)
- **Analytics:** Do you have proper tracking in place? Do you know which metrics tie directly to your goals? Seek alignment—metrics and objectives.

This audit gives you a baseline—and reveals what's already working and where your energy (and budget) should go next.

Step 3: Define Your Messaging and Value Proposition

Now that you know your goals and current standing, it's time to define how you want to be perceived. Ask yourself:

- What makes your brand unique?
- What core problem do you solve better than the competition? Remember, you're here to alleviate the consumer's Pain Points and solve problems.
- What tone and style will resonate most with your audience? Emotion is an effective tool, but use it wisely.

Keep it simple. Your value proposition should be short, clear, and emotionally engaging. This is the foundation of all your content—from website copy to social media captions to video ads. If people can't immediately understand the "why you," they won't stick around to find out.

It's often helpful to build a Buyer Persona-- A buyer persona is a fictional representation of your ideal customer. Fictional, yes, but based on a true story. The typical Buyer

Persona includes: age, gender, occupation, interests, and Pain Points. You create a buyer persona to better relate to and understand your target market. It helps you truly tailor your marketing efforts to their needs.

Create 2–3 detailed profiles representing your ideal customers. Include:

- Demographics (age, location, job title)
- Pain Points
- Motivations
- Decision-making behavior

These personas guide tone, channels, and content.

Step 4: Choose the Right Channels

Don't fall into the trap of being everywhere at once. Be where it matters. Use your audit and goals to identify the best platforms:

- **Search Engines (Google, Bing)**: Still the kings of purchase intent.
- **Social Media (Facebook, Instagram, LinkedIn, etc.)**: Ideal for engagement, brand awareness, and nurturing.
- **Email**: Often your highest ROI channel—especially for nurturing.
- **Paid Ads**: Great for targeting and speed, but require careful ROI management.
- **Content Marketing**: Long-term SEO and trust-building tool.

Ask: Where does your audience *live*, *scroll*, and *search*? Go there—and meet them with messaging that matches the platform's vibe…addresses Pain Points and offers easy to reach solutions!

Step 5: Map the Customer Journey

Every great digital strategy accounts for the **entire journey**—from "I've never heard of you" to "take my money." Think in phases:

- **Awareness**: Blogs, social media posts, YouTube videos
- **Consideration**: Webinars, whitepapers, case studies, product pages
- **Conversion**: Free trials, strong CTAs, landing pages, retargeting ads
- **Retention**: Email campaigns, exclusive offers, value-packed content
- **Referral**: Loyalty programs, user-generated content, reviews

Plot how people will discover you, trust you, buy from you, and return. This is your marketing map—and it guides everything else.

If you want to think of prospects in terms of the traditional sales or marketing funnel, organize your strategy into three core stages:

1. **Top of Funnel (TOFU)** – Awareness
   - Blog posts, SEO, social media, paid reach campaigns
2. **Middle of Funnel (MOFU)** – Consideration

- o Lead magnets, webinars, comparison guides, remarketing
3. **Bottom of Funnel (BOFU)** – Conversion
   - o Product demos, testimonials, case studies, sales pages

Match content to each stage to nurture your audience from interest to action. In other words, from 'I've got a problem' to 'Sale!'

Step 6: Create a Content Plan

Content isn't king—it's **fuel**. Your messaging drives the voice, your journey maps the format, and now it's time to plan:

- **What content are you producing?** (Video, blogs, emails, etc.)
- **Who's creating it?** (In-house, freelancer, etc.)
- **How often are you publishing?**
- **How will you promote it?** (Organic vs. paid)

Use a calendar. Build campaigns around promotions, holidays, and industry cycles. And remember—**quality > quantity**. One strong piece of content can outperform ten weak ones. It can often be re-purposed and re-promoted time and time again.

Step 7: Set Up Tracking and KPIs

Before you hit "launch," make sure you're tracking the right data. Define your Key Performance Indicators (KPIs) and make sure tools are in place:

- **Google Analytics 4** for web activity
- **Meta Pixel** or other platform tracking for paid ads
- **UTM parameters** to track campaign sources
- **CRM or Email tools** for customer behavior
- **Heatmaps or session recordings** for behavior insights

Execute and Test! Start launching content and campaigns with regular reviews:

- Monitor key metrics (traffic, conversions, engagement)
- A/B test where applicable
- Be agile—pivot based on performance

Your strategy is only as smart as your data. Don't fly blind—track everything. Just don't let vanity metrics distract from what truly matters: sales, leads, trust, and retention.

Step 8: Launch, Measure, and Optimize

It's go time—but launching doesn't mean "set it and forget it." Your digital strategy should be a **living, breathing system**. Once it's in motion:

- **Measure performance weekly and monthly**
- **Look for trends—not just outliers**

- **Tweak based on real feedback and results**
- **Double down on what works**
- **Cut or fix what doesn't**

Marketing isn't one-and-done. It is grounded in: test, learn, and evolve. The best marketers aren't just creative—they're curious, analytical, and humble enough to pivot when needed. Check your ego at the office of your bankruptcy attorney and focus on what is proven to be effective. Your brand matters. Results matter. Sales matter.

YOUR DIGITAL STRATEGY BLUEPRINT

STEP 1
Clarify Your Goals
- Define objectives
- Set SMART KPIs

STEP 2
Conduct a Digital Audit
- Evaluate current efforts
- Identify strengths & gaps

STEP 3
Define Your Messaging & Value Proposition
- Craft key messages
- Highlight your USP

STEP 4
Choose the Right Channels
- Select platforms
- Prioritize based on goals

STEP 5
Map the Customer Journey
- Outline stages
- Align content to journey

STEP 6
Create a Content Plan
- Plan types & topics
- Develop a content calendar

STEP 7
Set Up Tracking & KPIs
- Implement analytics
- Define key metrics

STEP 8
Launch, Measure, and Optimize
- Track performance
- Refine and improve

Launch, Measure, and Optimize

Conclusion: Your Next Move

You made it. You've powered through keywords, climbed the analytics mountain, dodged SEO snake oil, survived a crash course in CTRs and bounce rates, and even met my collie Captain. If that's not a journey, I don't know what is.

More importantly, you now hold a complete digital marketing blueprint—one that's been built brick by brick with real-world insights, a bit of humor, and a whole lot of practical, no-fluff strategy. Whether you are the marketing lead for a bustling brand or a beginner plotting your first campaign from a laptop at the kitchen table, the tools, tactics, and mindset outlined here are for *you*.

Remember that digital marketing is never "set it and forget it." This isn't a crockpot—(although I do miss Mom's cheesy crockpot potatoes!) but we're not slow-cooking content and hoping for the best. This is an ongoing, living strategy that should evolve as your audience, market, and brand evolve. That's why we emphasized things like SMART goals, regular audits, email segmentation, and— yep—even avoiding sending "buy now" clickers to blog posts about your childhood cat (although no sane person would ever complain about that, right!!!).

Let your strategy breathe. Keep asking questions. Test (and re-test) your assumptions. Track what matters (hint: sales), but also listen to your gut—especially when your gut tells you your significant other's marketing idea *might* need an A/B test first.

Above all, stay human. It's easy to get lost in platforms, pixels, and performance metrics—but never forget: behind every click is a person. Build trust. Be helpful. And when in doubt, go back to the basics: What problem are you solving, and for whom?

"In the midst of chaos, there is also opportunity."
— Sun Tzu, *The Art of War*

Now go find your opportunity. And if you need a partner to help bring your strategy to life, we're here to help.

☞ Visit **www.BrandVisionMarketing.com** to connect or just say hi!

You now have the structure, insights, and tools to launch or level-up your digital marketing.

You've got the blueprint to make it all happen.

Now go build something great…

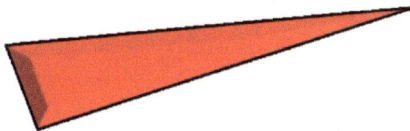

About BrandVision Marketing BrandVision Marketing is a full-service marketing agency dedicated to helping brands scale with clarity, creativity, and conversion in mind. Learn more or connect with us at brandvisionmarketing.com or visit or blog at brandvisionmarketing.wordpress.com.

VM: (865) 531-5874

EMail: info@brandvisionmarketing.com

Sources and Citations

- HubSpot (State of Marketing Reports)
- Google Analytics Documentation
- Moz SEO Guides
- Content Marketing Institute
- BrandVision Marketing Blog: brandvisionmarketing.wordpress.com
- Minecheck.com
- **1. Over 60% of shoppers begin on mobile devices**
  According to Google's *Think with Google*, shoppers initiate their online journey on mobile in **over 60% of shopping occasions** jmango360.com+15thinkwithgoogle.com+15linkedin.com+15.
- **2. 77% of retail consumers use mobile to research products**
  A study from Invoca found **77%** of retail shoppers use a mobile device to look up product details before buying invoca.com.
- **3. In eCommerce, more than 70% of traffic is mobile**
  OuterBox reports that mobile accounts for **over 70%** of eCommerce web traffic—and more than 60% of all website visits outerboxdesign.com.
- **4. Over 60% of B2B decision-makers start researching via mobile**
  On LinkedIn, a CMO expert noted that **more than 60%** of B2B decision-makers initiate their research on smartphones or tablets convinceandconvert.com+7linkedin.com+7bcg.com+7.
- Ahrefs

📘 List of Definitions: Digital Marketing Terms

A/B Testing (Split Testing):
A method of comparing two versions of content (like emails or ads) to see which performs better by testing one variable at a time.

Bounce Rate:
The percentage of visitors who land on your site and leave without taking further action.

Call to Action (CTA):
A prompt that tells the audience what to do next—like "Buy Now," "Sign Up," or "Learn More."

Click-Through Rate (CTR):
The percentage of people who click on a link or ad after seeing it.

Conversion Rate:
The percentage of users who complete a desired action, such as filling out a form or making a purchase.

Cost Per Click (CPC):
The amount you pay each time someone clicks your ad.

Cost Per Lead (CPL):
How much it costs to generate one qualified lead through marketing efforts.

Customer Lifetime Value (CLV):
The total revenue a business expects to earn from a customer over the lifespan of their relationship.

Digital Marketing:
All marketing efforts that use the internet or electronic devices—including websites, email, social media, and search engines.

Engagement Rate:
How actively people interact with your content (likes, shares, comments, etc.).

Funnel (Sales Funnel):
The journey potential customers take from first learning about your brand to making a purchase.

Impressions:
The number of times your content or ad is displayed, regardless of whether it's clicked.

Landing Page:
A web page created specifically for a marketing campaign that directs visitors to take one focused action.

Organic Traffic:
Website traffic that comes from unpaid search engine results.

Pay-Per-Click (PPC):
A model of internet advertising where advertisers pay each time someone clicks their ad.

Return on Ad Spend (ROAS):
A metric that calculates how much revenue you earn for every dollar spent on advertising.

Retargeting:
Serving ads to people who previously interacted with your website or content but didn't convert.

Search Engine Optimization (SEO):
The process of optimizing your website to rank higher in search engine results.

Traffic Sources:
The origins of your website traffic—such as organic (search), direct (typed in), referral (from other websites), or paid (ads).

UTM Parameters:
Tracking codes added to URLs to help measure the effectiveness of online campaigns.